# Fixing Cars

Sharon Holt Adam Nickel

Australia • Brazil • Japan • Korea • Mexico • Singapore • Spain • United Kingdom • United States

**Fixing Cars**

**Fast Forward**
**Green Level 14**

Text: Sharon Holt
Illustrations: Adam Nickel
Editor: Johanna Rohan
Design: James Lowe
Series design: James Lowe
Production controller: Emma Hayes
Audio recordings: Juliet Hill, Picture Start
Spoken by: Matthew King and Abbe Holmes
Reprint: Siew Han Ong



ISBN 978 0 17 012587 1
ISBN 978 0 17 012585 7 (set)

**Cengage Learning Australia**
Level 7, 80 Dorcas Street
South Melbourne, Victoria Australia 3205
Phone: 1300 790 853

**Cengage Learning New Zealand**
Unit 4B Rosedale Office Park
331 Rosedale Road, Albany, North Shore NZ 0632
Phone: 0800 449 725

For learning solutions, visit **cengage.com.au**

Printed in Australia by Ligare Pty Ltd
6 7 8 9 10 11 12 20 19 18 17 16

Evaluated in independent research by staff from the Department of Language, Literacy and Arts Education at the University of Melbourne.

Sharon Holt Adam Nickel

## Contents

Chapter 1

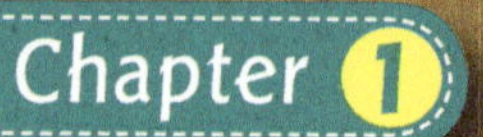

# Bonnie's Gift

Fixing cars was Bonnie's thing.
Her teacher called it a gift.
She said Bonnie was the only kid
she knew who could look
under the hood of a car
and name all the engine parts.

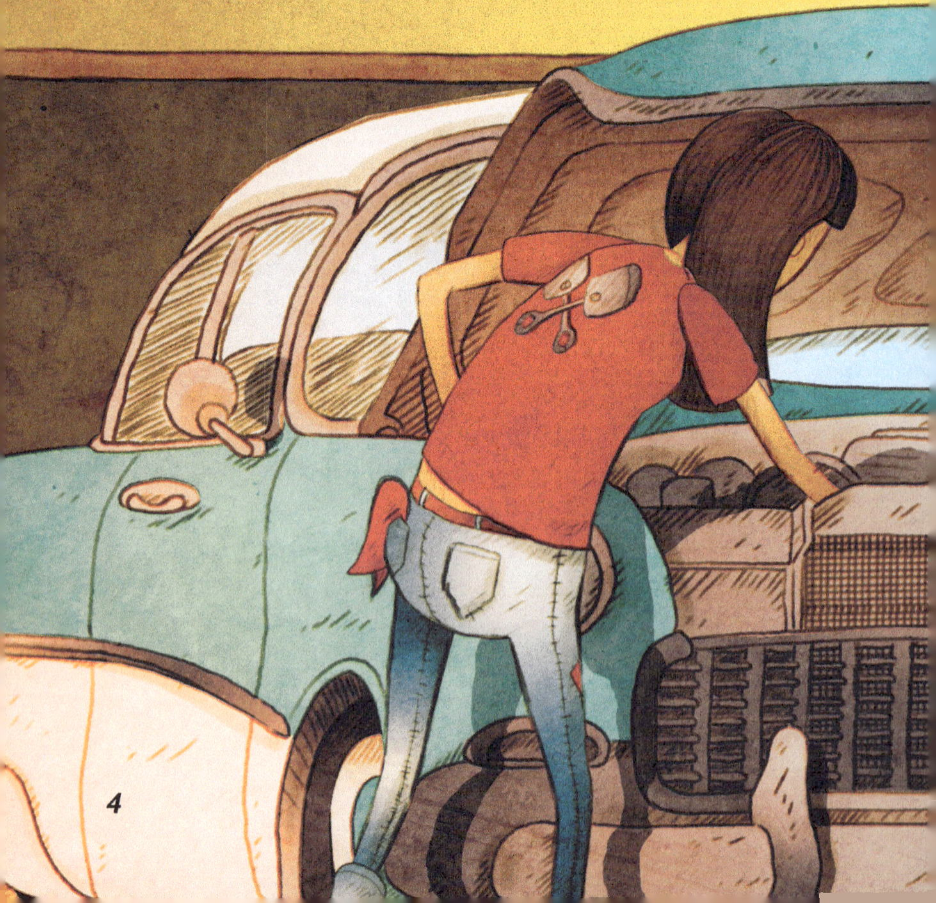

Bonnie's friend Rick shook his head as Bonnie changed the fan belt on her teacher's car.

"I don't know how you do it," said Rick.
"I can't even find the engine in my mum's van!"

"That's because it's under
the passenger seat,"
Bonnie laughed.
"Come on.
We don't want to miss the bus."

"I'm looking forward to the day when we don't have to take the bus," said Rick.
"Will your dad buy you a car when you learn to drive?"

"You really don't know my parents very well, do you?"
Bonnie laughed.

Chapter 2

# The Problem at Home

Bonnie's gift for fixing cars might have looked good to her teacher. But, her parents had a very different idea.

They believed girls should learn
to cook and clean.
They always gave her dolls
and soft toys for her birthday.
But, all Bonnie ever wanted
were model cars.

Running Words 176

"Cars!" said her mum.
"What use are model cars?
Girls need nice dresses
and jewellery!"

Luckily for Bonnie,
her grandpa understood.
He always gave Bonnie model cars
for her birthday.
In the school holidays,
they worked on her grandpa's old car
every day.

"I love spending time with you, Grandpa," said Bonnie. "You don't make me stand over a hot frying pan."

"No," said Grandpa.
"We like standing over hot engines much better!"

Chapter 3

# The Car Trip

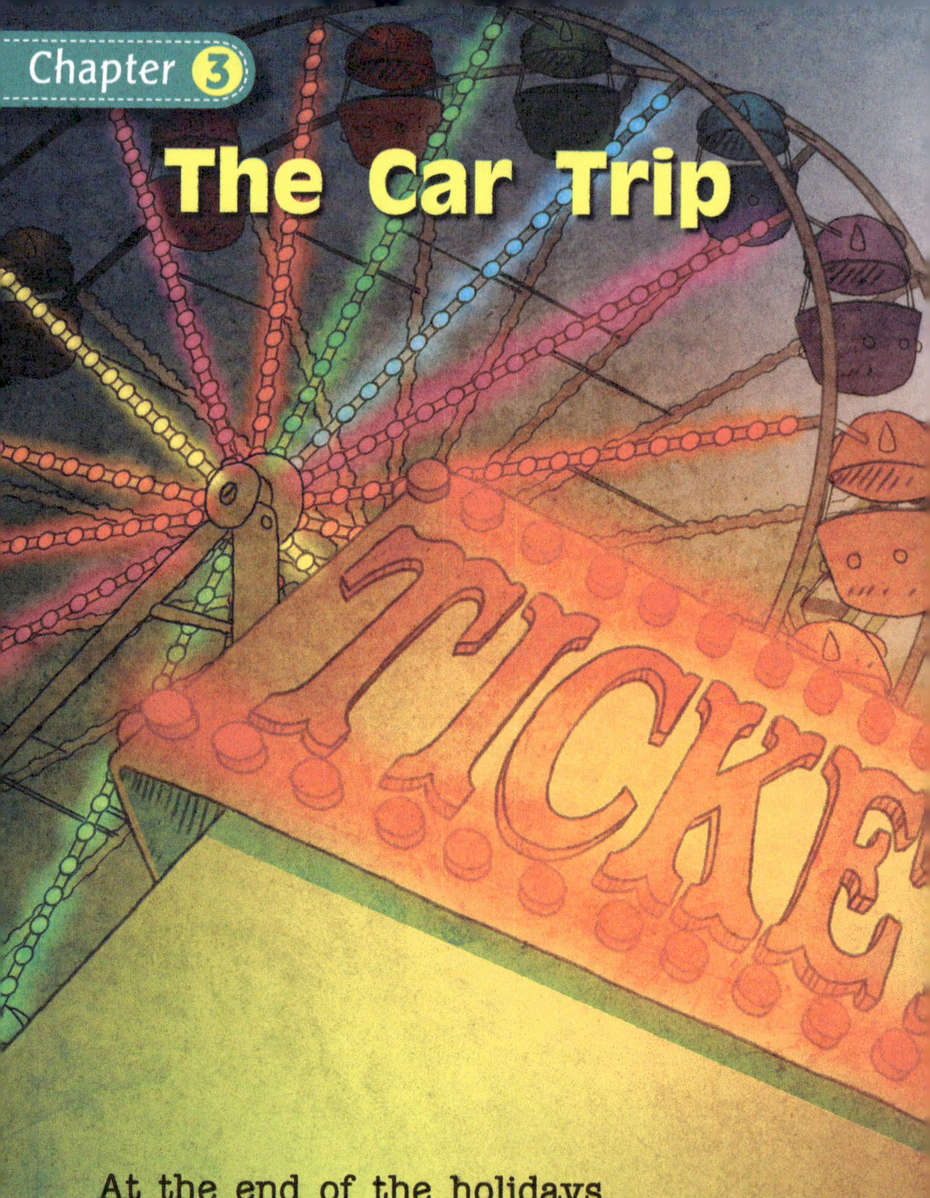

At the end of the holidays, Bonnie's parents took her to a fun park with a giant roller coaster.

Bonnie was excited about the trip. She knew that a giant roller coaster would have a very big engine.

On the way,
Bonnie heard a strange sound
coming from under the hood.
"Hey, Dad," she said.
"Can you hear that noise?
It sounds like the fan belt."

Dad stopped the car.

"Don't worry," said Bonnie.
"I can fix it."

She jumped out of the car
just as rain started to fall.

Bonnie's mum watched Bonnie get soaked as she grabbed her tools from the back of the car.

"Come in out of the rain," Mum shouted. "Your hair looks a mess."

Bonnie's dad laughed. "I think she likes it that way," he said.

Before long, Bonnie slammed the hood and got back into the car.
"The fan belt was loose," she said.
"It should be okay now."

"How did you know how to fix the fan belt?" asked Mum.

Bonnie smiled.
"It was nothing really.
Just something Grandpa showed me."

“I thought you helped Grandpa in the kitchen,” said Mum.

“I did,” said Bonnie.
“It’s just that I helped him in the garage first.”

Bonnie’s dad laughed.
“I guess playing around with cars can be useful after all!”